Plant
Tricksters

Plant Tricksters

Janet Halfmann

Franklin Watts
A Division of Scholastic Inc.
New York • Toronto • London • Auckland • Sydney
Mexico City • New Delhi • Hong Kong
Danbury, Connecticut

To my farmer dad, who inspired in me a love of all of nature

Note to readers: Definitions for words in **bold** can be found in the Glossary at the back of this book.

Photographs © 2003: Animals Animals: 20 (Deni Bown/OSF/Earth Scenes), 37 (John Pontier); Bruce Coleman Inc.: 36 (Jean-Claude Carton), 6 (A.J. Deane), 40 (Edward R. Degginger), 46 (Kenneth Fink), 47 (Wayne Lankinen), 31 (James Simon); Carr Botanical Consultation/Gerald D. Carr: 52; Corbis Images/Kevin Schafer: 12; Dembinsky Photo Assoc./Skip Moody: 48; Nature Picture Library Ltd.: 5 right, 24 (Georgette Dowma), 10 (Neil Lucas), 39 (William Osborn); Peter Arnold Inc./Darlyne A. Murawski: 13; Photo Researchers, NY: 2 (Geoff Bryant), 28 (Ray Coleman), 16 (E.R. Degginger), 27 (John Eastcott/ Yva Momatiuk), 42 (R.J. Erwin), cover (Fletcher & Baylis), 22, 49 (Kjell Sandved), 45 (Sercomi/BSIP/SS), 5 left, 19 (Jim Zipp); Shoko Sakai: 23; Visuals Unlimited: 51 (R. Calentine), 21 (Ray Coleman), 9 (Wally Eberhart), 30 (Dick Keen), 32 (Genn M. Oliver), 29 (Kjell B. Sandved), 15 (John N. Trager), 35 (Tom Uhlman).

The photograph opposite the title page shows the large bloom of the pelican flower, which attracts flies with its foul odor.

Library of Congress Cataloging-in-Publication Data

Halfmann, Janet.
    Plant tricksters / Janet Halfmann.
       p. cm. — (Watts library)
    Summary: Introduces various plants that use unusual defense mechanisms to survive.
    Includes bibliographical references (p.   ).
    ISBN 0-531-12278-6 (lib. bdg.)   0-531-16371-7 (pbk.)
    1. Plants—Insect resistance—Juvenile literature. 2. Plant defenses—Juvenile literature. [1. Plant defenses. 2. Plants—Disease and pest resistance.] I. Title. II. Series.
SB933.2 H35 2003
581.4'7—dc21

2002015338

Contents

The beautiful mirror orchid looks and smells so much like a female wasp that males try to mate with it, thus pollinating the flower.

Chapter One

The Imposters

Plants lead busy lives. Every day, they need to take in food and water. They must constantly battle armies of nibblers, sippers, and munchers. They have to find mates in order to make **seeds**. And those seeds must travel to a suitable place to grow into new plants.

For plants, as for animals, every day is a struggle to survive. But unlike animals, plants cannot run or crawl or fly. They are rooted in one place. Over millions of years, plants have been **adapting** in countless ways to help them survive in their surroundings. Animals, which depend on plants for food, shelter, and other things, have been changing along with them.

Perhaps the biggest adaptation of all in the plant world was the development of flowers about 125 million years ago. Early plants, such as mosses and ferns, did not have flowers. They had to rely mainly on wind and water to spread their **spores**.

Plants with flowers, however, can attract more efficient helpers—the animals—to spread their **pollen**. Flowers contain a rich food supply of **nectar** and pollen. The flowers advertise this feast with a wide variety of colors, odors, and shapes. Insects and other hungry animals come to eat, and in the process become the plants' pollen delivery service when they travel from plant to plant to eat.

But sometimes, a plant stands a better chance of surviving and reproducing if it doesn't look like a plant. We are all familiar with animals that aren't what they seem. For example, an inchworm can look like a tiny branch and escape being eaten by a hungry bird. Plants, too, sometimes have looks that can fool.

The "Insect Orchids"

Some of the most beautiful and unique flowers are in the orchid family. The orchids lure a wide variety of insects to carry their pollen from flower to flower. While most flowers reward their animal messengers, called **pollinators**, with nectar or pollen, orchids rarely do. Some are imposters, mimicking female bees, wasps, flies, and other insects.

How Is a Flower Pollinated?

A flower is a container for the male and female parts of a plant. Often, both are in the same flower, at its center. The male part is a **stamen**. It consists of a slender stalk called a **filament** with a swollen **anther** at the top that releases male cells called pollen. Generally, several stamens ring the female part, called a **pistil**. The pistil consists of a stalk called a **style** with a sticky **stigma** at its tip for receiving pollen and an ovary containing **ovules** at the bottom.

Pollination occurs when pollen sticks to the stigma of a flower. To be successful, most plants require that the pollen and the stigma be from different plants. This is called **cross-pollination**. When pollen lands

on the right stigma, it sends a long tube down to the ovule to **fertilize** it. The fertilized ovule develops into a seed, which is the plant's offspring.

The mirror orchid (*Ophrys speculum*) is a common insect **mimic** that grows in Europe. Its lower petal is shaped like an oval mirror. The mirror petal glistens a metallic violet-blue, like the wings of a wasp. Long reddish hairs fringe the petal's yellow edge, copying a wasp's furry body. Two side petals look like insect wings. To add to the illusion, the flower smells like a female wasp!

Orchids that mimic insects bloom early in spring before female insects appear. Then the fakes don't have to compete with the real thing.

The orchid's perfume lures male wasps in search of a mate. A male tries to mate with the fake female, but instead gets two packets of pollen stuck like horns to his head. The trick works so well that the male soon is fooled again. He smells another mirror orchid and tries to mate with it. Instead, he delivers his pollen cargo, bringing about the pollination of the mirror orchid.

The hammer orchids (*Drakaea*) of Australia are even more intriguing. Their flowers mimic not only the shape and smell

Elbow orchids, like hammer orchids, mimic wingless female wasps. The petal that looks like the fake wasp is to the left.

of a female wasp that lives in Australia, but also her behavior. The real female wasp doesn't have wings, so she climbs to the top of a plant when she is ready to mate. There she waits for a male to come by and carry her off.

A petal of the hammer orchid mimics this waiting female. The fake is maroon-colored and seems to be hanging from a stem, just like the real female. Before long, a male lands and tries to fly off with the fake, but she won't budge. The male's wild motions make the petal swing forward, somersaulting him head first into the flower. Two pollen packets stick to his back. He's been tricked!

Fake Flocks of Aphids

The showy Rothschild's slipper orchid (*Paphiopedilum rothschildianum*) grows only in Southeast Asia. The flowers bloom in clusters and have bizarre coloring—greenish-yellow and rose with maroon stripes. Each flower has a slipper-shaped cup in the center and long, thin, winglike petals stretching out on each side.

The spicy smell of the flowers attracts female hoverflies, plump with eggs. These flies usually lay their eggs on a plant among a colony of tiny insects called **aphids**. The aphids provide food for the larvae after they hatch.

No aphids live on the slipper orchid, but that's not the way the flies see it. This orchid trickster has tiny markings that look like a flock of aphids. Female flies land and lay their eggs among the fake aphids. Some of the flies fall into the orchid's

Marks on this Asian slipper orchid look like aphids to attract hoverflies that lay their eggs among the tiny insects.

cup and become trapped. The only way out is through two exits near the flower's male and female parts. As the flies escape, they deposit pollen they are carrying from other slipper orchids and pick up a new load.

The Passionflower Vine's Fake Eggs

Why would a plant make fake eggs? Just tag along with the colorful female *Heliconius* butterfly and you'll find out. This butterfly of warm forests of the Americas always lays her eggs on a passionflower vine (*Passiflora*). It's the only plant the caterpillars that hatch from the eggs will eat. They eat a lot, which isn't good news for the passionflower vine.

Often, as the female butterfly checks out a passionflower vine, she finds bright yellow little knobs on its leaves. They look like eggs laid by another butterfly, but they are really fakes made by the plant. The trick works. The mother butter-

Tiny yellow knobs on this passionflower vine are fake eggs made by the plant. They may influence the butterfly to lay her eggs elsewhere.

More Tricks

Some passionflower vines use another trick to keep from being devoured by baby caterpillars. The vines produce nectar to attract ants onto their leaves; then the ants eat the butterfly caterpillars.

fly flies away to lay her eggs elsewhere. That way, her babies won't have to compete for food. The plant is safe—for now.

Stone Plants

In a desert, fat leaves filled with water are bound to attract thirsty animals. But in Africa, animals walk right past the plump leaves of the stone plants (*Lithops*). Why? Because the plants look exactly like stones, even growing among real stones and rocks.

Known as "living stones," these plants have two round, fleshy leaves, with a groove between them. Lines, dots, or blotches often mark the top of the leaves. The markings and colors of the plants match the rocks and pebbles where they grow. Some are gray or blue; others yellow, orange, brown, or even white. See-through areas at the tips of the leaves act like windows to let in light for making food. The stems of the plants are underground.

Only when it rains in the fall does a large yellow or white flower rise out of the groove between a plant's leaves. The flower makes the plant visible, but because the rain provides plenty of water for the animals, they leave the plant alone. The flower quickly makes and spreads its seeds. By the time the desert dries out, the plant looks like a stone once again.

Animal Nicknames

Stone plants look like tiny hoofprints of cattle, so farmers in Africa call the plant cattle hoof or sheep foot.

Desert animals don't bother these water-filled stone plants because they look like stones. The flower appears during the rainy season.

The skunky smell of skunk cabbage wafts through the air in early spring, attracting the first insects of the season.

Chapter Two

Rotten Smells

When we think of flower scents, the odor of sweet perfume comes to mind. In fact, a huge perfume industry is built on flower scents. But flower scents are not meant to please us. They exist to attract animal pollinators.

Bees, butterflies, and moths favor sweet fragrances just as we do. But for many flies, beetles, and gnats, nothing beckons like the foul smell of rotting meat or animal droppings. To these insects, the foul odors signal a meal or a perfect place to lay their eggs. Many flowers imitate the look and smell of rotting flesh or animal droppings. The insects swarm to the stinky flowers, but find nothing to eat, and often a trap instead.

The Skunk Cabbage's Warming Hut

The first insects out and about at the end of winter smell skunk cabbage blooming in wet areas of eastern North America and Asia. Skunk cabbage (*Symplocarpus foetidus*) belongs to the arum family, a plant group with lots of stinky flowers.

The skunk cabbage's stalk of tiny, yellow flowers, wrapped inside a purplish hood, pushes up while the ground is still frozen. The plant produces so much heat that it melts the ice and snow around it. Inside the purplish hood, the air stays a cozy 59 to 72 degrees Fahrenheit (15 to 22 degrees Celsius) during the two weeks the flowers bloom. Outside, the temperature may still be below freezing. The heat carries the skunky odor of the flower far and wide.

Insects, especially flies that lay their eggs on rotting meat, zoom toward the skunky odor. Inside the plant's hood, they find a warm, cozy place to mate and lay their eggs. During their stay, their bodies pick up pollen. Some of the flies carry

Check the Thermometer

Scientists have found that skunk cabbage produces more or less heat depending on the temperature outside. In cold weather, the plant produces more heat, and in warm weather less, thus keeping its temperature at the same cozy, constant level.

the pollen to another stinky skunk cabbage, and the plant is pollinated.

The Stinky Dead Horse Arum

On islands in the Mediterranean, blowflies can't resist the rotting stench of the dead horse arum (*Helicodiceros muscivorus*). This plant often grows in the middle of a colony of breeding gulls. Among dead fish and gull droppings, the plant spreads out what looks like a grayish-purple dinner plate. Streaked with veins and covered with dark red hairs, the new dish looks and smells like a dead animal. Blowflies land on it and crawl into the smelly hole at its center searching for food. Once inside, they can't get out.

Bees and other insects enter the warm, cozy hood of the skunk cabbage. As they mill around, they pollinate the tiny flowers inside.

Blowflies mistake the dead horse arum for a rotting animal. They crawl into its dark hole and are trapped while they pollinate the flowers inside.

Inside are the plant's flowers, male flowers at the top and female flowers at the bottom. The flies crawl down to the female flowers, which provide nectar for them. As they drink, some of the flies deposit pollen brought from other dead horse arums. The warm, dark hole seems perfect for laying eggs, and many female blowflies do just that. But any maggots that hatch will die because there is nothing for them to eat.

A day or two later, the female flowers shut off the nectar supply. The male flowers above them now shed their pollen, covering the trapped flies. The hairs at the entrance wither and the flies escape. The stench of another newly opened dead horse arum fools some of the flies, and they deliver their pollen load.

Sliding into a Dutchman's Pipe

Dutchman's pipe (*Aristolochia*), which grows mainly in warm regions, has another kind of trap. Its foul-smelling flowers have a tube leading to a fat, pipelike chamber. When a flower opens, the tube points upward. Its entrance flares out like a welcoming flag. But flies attracted by the flower's foul smell find themselves on a slippery slide. They slip down the waxy tube into the pipelike chamber. Long, stiff hairs inside the flower keep them from escaping.

Inside their prison, a false window near the flower's stigmas lets in light. The tricked flies zoom toward the fake window to try to escape. There they find nectar. They stop to eat and leave behind any pollen they have brought. A few days later, the flower's anthers open and shower the flies with new pollen. Now, the prison hairs wilt and the flower bends sideways, letting the flies walk out. They soon smell the foul odor of another Dutchman's pipe and slide into another prison.

Blooming Starfish

In deserts of Africa, strange star-shaped flowers bloom on cactuslike plants. Called starfish flowers (*Stapelia*), they belong to the milkweed family, another group with lots of stinkers. Many

Flies attracted by the foul smell of the Dutchman's pipe soon slide down its slippery tube into a trap.

Long Traps

The fishy-smelling pelican flower (*Aristolochia grandiflora*), a cousin of the Dutchman's pipe, has flowers more than 3.3 feet (1 meter) long.

Smelly flowers shaped like starfish bloom in deserts of Africa. Female flies often lay their eggs on the flowers.

starfish flowers are large—up to 16 inches (41 centimeters) across.

Starfish flowers are usually flesh colored or dark purplish-red. Hairs often fringe or cover the fleshy petals, resembling mold or fur. The leathery flowers stay open for days and smell like rotting meat or fish. To flies, the flowers appear to be dead, furry animals. Female flies fooled by the flowers often

Moving Lures

Some milkweed flowers, such as *Ceropegia*, have special hairs or knobs that move in the wind to lure flies to their flowers.

deposit their eggs or maggots on them. But the young die because there is no food to eat.

"Purple Dung Flower"

Dung beetles can't resist the tantalizing smell of animal **dung**. They fly to the aroma for a tasty meal or to lay their eggs. A beautiful purple flower (*Orchidantha inouei*) that grows near the ground in Borneo gives off a smell just like dung.

The flower, related to bananas and gingers, attracts a kind of small beetle that tunnels under dung piles. The beetles don't find any dung in the flower, or even any nectar. But as they crawl around under the petals, they deposit pollen from other purple dung flowers. They exit the flower with a new supply of white pollen stuck to their backs. As the beetles zoom to other "purple dung flowers," they pollinate them.

This pretty purple flower of Borneo attracts dung beetles. The flower smells like dung or animal droppings — the beetles' favorite food.

*Some kinds of
Bauhinia flowers
change color after
they are pollinated.
An entire flower may
shift color, or just
parts may change.*

Chapter Three

Color-Change Artists

The colors of flowers are like flashy restaurant signs. They lure animal pollinators to flowers by signaling that a rich supply of food is ready for them. Different animals come, depending on the color of the flowers. In general, bees prefer yellow, blue, and purple flowers, butterflies bright colors or white, birds red or orange, and moths and bats light-colored flowers that open at night.

The flowers of more than 450 **species** of plants are color-change artists. They

Many flowers have lines, dots, or patches of contrasting color called **nectar** or **honey guides**. They guide pollinators to nectar or pollen much like the runway lights at an airport. Many of these marks are in **ultraviolet**, which is visible to insects but not to people. After pollination, some nectar guides change color, such as in horse chestnut tree flowers and some blue lupines.

change color after they are pollinated and their food supply decreases. This means the plant can make more seeds. An entire flower may shift color, or one or more parts that are highly visible to a pollinator may change. In still another kind of color change, a patch of wildflowers in Arizona shifts color with the seasons as its pollinator changes.

Yellow Lantana Magic

A thicket of yellow lantana (*Lantana camara*) bushes puts on a color magic show. The clusters of tiny flowers on the bushes wear a trio of colors. On opening, each tiny flower is yellow and contains nectar. Butterflies visit to drink and, in the process, carry pollen from one yellow flower to another. Once a yellow flower is pollinated, it turns orange and finally red.

Scientists say that the color change signals to butterflies that a flower no longer contains nectar. In experiments, butterflies visited yellow flowers much more frequently than orange or red ones. Yellow lantana's color magic benefits both the flowers and the butterflies. The flowers gain because butterflies visit only blooms that need pollinating and don't waste

time on those already pollinated. The butterflies benefit by visiting only flowers that contain nectar.

Disappearing into the Night

Moths and bats feed at dusk and during the night. They are attracted by white and light-colored flowers that stand out against the darkness. After pollination, several of these night-blooming flowers change to darker colors. Then, presto! They seem to disappear into the night.

For example, flowers of the moth-pollinated rangoon creeper (*Quisqualis indica*) open white at dusk. By the following evening, they are dark red. Similarly, flowers of the bat-pollinated angel's tears (*Brugmansia versicolor*) open creamy

Yellow lantana flowers start out yellow, then turn orange and finally red. Scientists say the color changes are a signal to butterflies.

white and then turn reddish-orange. The darker flowers can't be seen by the pollinators because they blend in with the dark night background. Again, the color magic means more flowers are pollinated, and the pollinators feast on more nectar.

Overnight in an Amazon Water Lily

The huge flower of the Amazon water lily (*Victoria amazonica*) opens for the first time in the evening. The creamy white flower is warm and has the sweet smell of butterscotch mixed with pineapple. The perfume attracts swarms of large beetles, related to June bugs. They settle on the center of the flower to feast on small knobs of sugar and starch. As they eat, some of the beetles leave pollen from visits to other flowers on the lily's stigmas. During the night, the petals slowly close around the beetles, trapping them inside. They remain trapped all the next day, eating and getting very sticky.

Toward evening, the lily's anthers shower the beetles with pollen and the flower reopens. It is now pink and has lost most of its scent. The beetles, covered in pollen, are free to go. Pink flowers don't attract them, so they fly straight to a fragrant white flower on another plant. Soon, they get trapped again and deposit their pollen on the new flower.

The pink bloom the beetles left behind closes and sinks into the Amazon River. The seeds it produces eventually float to the surface to be carried to a new home by the current.

When the giant flower of the Amazon water lily closes around its beetle pollinators it is white, but when it reopens it is pink.

Designer Leaves

The leaves of the Amazon water lily are so strong and buoyant that they can support a child. One plant can have fifty leaves, measuring up to 6.6 feet (2 m) across. Birds called lily trotters trot around on the leaves hunting for insects. The elaborate network of veins on the underside of the leaves inspired the design for the cast-iron supports of the Crystal Palace of 1851 in London.

Red in Summer, White in Fall

Scarlet gilia (*Gilia aggregata*), or skyrocket, is a wildflower known for its bright red, trumpet-shaped flowers. But on Fern Mountain in Arizona, the plant's flowers aren't always red. They range from red to pink to white. Early in the summer, most of the flowers are red. As the season advances, the flowers become lighter. By late August, most of the blooms are white.

Scientists have found that this shift in color coincides with a change in the flower's pollinators. In early summer, hummingbirds are the main pollinator of scarlet gilia. They prefer

red flowers, and that's what they find. But in late summer, the hummingbirds leave Fern Mountain. That leaves the hawk-moth as scarlet gilia's only pollinator. Hawkmoths fly at night. They can see white flowers better, and that's what they find. By shifting color as its pollinator changes, more scarlet gilia are able to produce seeds.

On Fern Mountain in Arizona, scarlet gilia changes from red to white as its pollinator switches from hummingbirds to moths.

The sensitive plant folds up its leaflets at the slightest touch. Animal nibblers are left with nothing but stems.

Chapter Four

Battling Robbers and Munchers

Plants provide most of the food for all animal life on Earth, including people. Some animals eat plants directly, the rest eat animals that eat plants. That means plants are constantly under attack. Because plants can't run away, they have had to develop an amazing arsenal of defenses—sharp spines, traps, poisons, and some very tricky devices.

The Sensitive Plant

Plants constantly face a dilemma. They must stretch their leaves out wide to catch sunlight to make food, but then they are easy targets for munchers. The feathery leaves of the sensitive plant (*Mimosa pudica*) fan out in hot regions of America, Asia, and Africa. Grasshoppers and locusts hop on the plants, expecting a delicious meal.

But a hungry insect gets a surprise. When it touches just one **leaflet**, all of the plant's leaves fold up. They collapse one after another, like a stack of falling dominoes. Within seconds, only stems are left. If the insect still doesn't give up, the plant moves again, exposing sharp spines.

The sensitive plant's leaves fold up and droop because of a decrease in water pressure. The plant can pump its leaves back up in about twenty minutes.

Fake Weapons

The leaves of the deadnettle (*Lamium*) look like those of the stinging nettle, but don't sting. Nibblers leave the copycat plant alone.

Stinging Nettle

True to its name, the stinging nettle (*Urtica dioica*) stings! Hairs on its stem and leaves work just like the needles doctors use to give injections. Each hair is a tiny, hollow, glassy spike. At the slightest touch, the tip breaks off at a slant. This sharp, broken edge can cut an animal's skin. Then poison at the bottom of the hair squirts into the wound.

The nettle's sting keeps rabbits and other large animals away. But small animals, such as the caterpillar of the red admiral butterfly, can easily chew around the needles without being stung.

Rattan Homes for Ants

Plants around the world team up with animal bodyguards. Several kinds of rattans (*Calamus*) have ant defenders. Rattans are spiny, climbing palms that cling to trees and other plants in the rain forests of Asia.

Rattan stems often grow extremely long. But they grow only from the bud at the tip. These tips are sweet and tender, so they are a favorite of squirrels, wild pigs, tapirs, and other animals. Sharp spines near the tips help keep nibblers away. But many rattans also have extra-roomy leaf husks or other spaces near the tips that make perfect homes for ants.

The thousands of bristly hairs of the stinging nettle stop most big animal nibblers, but little ones can chew around the stingers.

Many kinds of rattans provide special nesting spaces for ants. In return, the ants defend the rattans against hungry animals.

The ants noisily defend their home. If an animal touches the rattan, the ants beat their heads against the leaf husks. Soon, thousands of angry ants are beating their heads in unison. They sound like a hissing snake. If the nibbler persists, the ferocious ants swarm and bite. The ant bodyguards help keep the rattan safe.

Snapdragon Jaws

All kinds of insects visit flowers, but not all of them make good pollinators. Ants, for example, are often robbers. They are so small and smooth that they can feast on flower nectar without picking up any pollen. Many flowers have developed ways to keep out such robbers.

The snapdragon (*Antirrhinum*), a popular garden flower, is one of them. Its petals tightly block the entrance to the flower, like the jaws of a dragon. Only big, heavy bumblebees are strong enough to force open the petal jaws.

The bumblebee pushes its head into the flower's deep tube and drinks the nectar with its long tongue. As it feeds, it delivers pollen from other snapdragons and collects a new supply.

Flower Toy

People of all ages enjoy pinching open the snapdragon's dragon jaws and watching them snap shut.

The jaws of this snapdragon don't let in just any insect. Only big, heavy bumblebees are strong enough to force apart the petals.

The flower's scent also rubs off on the bumblebee. It carries the smell back to the hive and more bumblebees visit the plant.

The Daisy's Poison

Plants are the original makers of **insecticides**. Some carry poisons in their flowers, others in their leaves, roots, or seeds. Daisies (*Leucanthemum*) and other flowers in the sunflower family make some of the strongest poisons of all.

A meal of daisy flowers doesn't make an armyworm feel too bad at first. But as the hot sun shines on the armyworm, it feels sicker and sicker. That's because the power of the daisy's poison multiplies 10 to 1,000 times when exposed to sunlight. The armyworm literally "toasts" in the sun.

But even such a powerful poison doesn't keep the daisy completely safe. A hungry caterpillar called a leaf roller bends the daisy's petals over itself and ties them with silk. Shaded from the sun, it can eat unharmed by the plant's poison.

This sunny meadow of oxeye daisies looks inviting, but insects beware. Daisies make some of the most powerful insect poisons.

The Potato's On-and-Off Switch

In some plants, defenses turn on only as needed. An example is the potato plant. When an insect nibbles on the leaf of a potato plant, a chemical is produced that interferes with the insect's digestion. The poison makes the animal nibbler grow more slowly. Then the insect's enemies are more likely to attack it.

This corn-eating caterpillar may soon be under attack. Corn plants being eaten can send out odors to call wasps to their rescue.

The nibbled leaf also sends out a chemical signal that turns on the defenses of nearby potato plants before the insects reach them. Scientists have found that plants sometimes even warn plants different from themselves. For example, sage plants can signal tomato plants.

Corn SOS

In the war against munching animals, some plants call other animals to their rescue. When corn, cotton, and other crops are under attack by caterpillars, they send odor signals to wasp enemies of the caterpillars. The wasps come and deposit their eggs inside the caterpillars. When the eggs hatch, the larvae devour the caterpillars.

Scientists have found that each plant's signal is specific, aimed at only one kind of wasp. The plant can tell what kind of caterpillar is eating it by the taste of the insect's spit. Then the plant tailors its odor signal to appeal only to the wasp that lays its eggs inside that kind of caterpillar.

Plant Insecticides

The poisons of many plants, such as the pyrethrum daisy, are deadly to insects, but not to people and other **mammals**. Some of these plant-based poisons can be used as insecticides.

In spring, the male cones of the pine tree release clouds of pollen. The wind carries the pollen grains to female pinecones.

Chapter Five

Grab Bag of Tricks

Pretend that you are a plant. Try to think of a clever way you could spread your pollen or seeds, get water, or battle enemies. Chances are a plant somewhere is already using your tactics.

Flying with Balloon Wings

Before plants developed flowers, pollen was spread mainly by the wind, and occasionally by water. Many plants today, including most trees and grasses, still rely on the wind. Wind-pollinated plants need to produce millions of pollen grains because most miss their target. All pollen is tiny, like specks of dust. Pollen carried by the wind tends to be the tiniest of all. It is also light and dry for smooth sailing through the sky.

Pine trees (*Pinus*), which dominate many of the world's forests, have been growing on Earth since long before the flowering plants. For millions of years, they have relied on the wind to carry their pollen. The pine's male and female parts are in cones. Both male and female cones grow on the same tree. In spring, the male cones release huge clouds of yellow pollen.

Pine pollen is large for a wind-carried pollen. But it has a special trick that makes it a champion traveler. Each pollen grain has two balloonlike wings to help it fly. The two small sacs filled with air look like mouse ears. Pine pollen can travel more than 500 miles (800 kilometers) and still be able to do its job. Because pine trees grow in thick stands, their pollen

Pollen Storytellers

Pollen grains have many stories to tell. Pollen grains are so tough that they can lie buried for tens of thousands of years and still be recognizable when dug up. In addition, the pollen of each kind of plant is as unique as a person's fingerprint. Scientists study pollen grains to find out which plants grew in a certain place during a particular time in history. The uniqueness of pollen grains also can provide clues at crime scenes. For example, if a dead body discovered outdoors has pollen on it from an area different from where it was found, police know the body was moved.

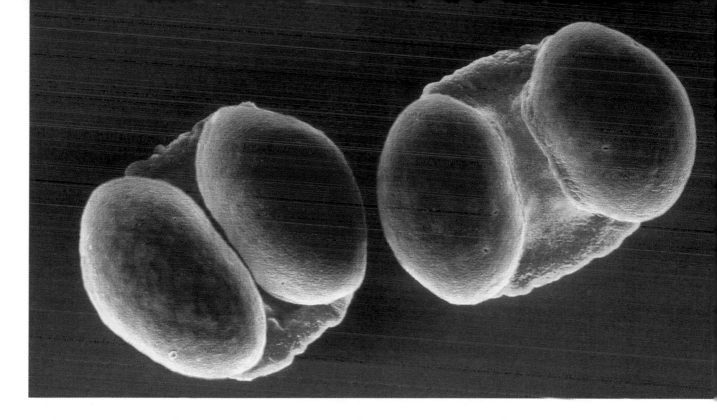

grains have a good chance of landing in the female cones of another pine tree. Female cones have a trick of their own. Each ovule has a sticky drop of liquid to catch a pollen grain.

Pine pollen is large for a wind-carried pollen. Two special sacs of air help the pollen fly as far as 500 miles (800 km).

Plants of the Dinosaurs

Cycads, ancient plants with palmlike leaves, shared Earth with the dinosaurs some 200 million years ago. Back then, cycads grew in huge forests and the dinosaurs nibbled on them. Today, many kinds of cycads are on the verge of extinction. Scientists long believed that cycads were wind pollinated, but they now know that many kinds attract weevil and beetle pollinators.

Cycads, like pines, have cones. But their male and female cones are on different plants. A Mexican cycad called the

Cycads were so abundant when the dinosaurs lived that this time is sometimes called the Age of the Dinosaurs and Cycads.

The cones of some cycads lure beetles or weevils. The insects mate in the male cones, then their young carry pollen to female cones.

cardboard "palm" lures the snout weevil to be its helper. The action starts with the male cone. When its pollen is about ready, it heats up and gives off an odor the weevils can't resist. The snout weevils come to the cone to eat, mate, and lay eggs. Their young grow to adults in about a week. The new weevils feed on starch in the cones and, in the process, are dusted with pollen.

Now, the new weevils fly to other cycad cones. The female cones don't have any food to offer, but they imitate the male cones that do. They heat up like the males and give off the same tantalizing smell. Some of the weevils are fooled. They visit the female cones and leave their pollen loads behind, making it possible for the cones to make seeds.

The Exploding Scotch Broom

In spring, Scotch broom (*Cytisus scoparius*) turns hillsides golden with bright yellow sweet pea-shaped flowers. Bumblebees zoom to the pollen-rich blooms. But the bees soon find out that the flowers are like little firecrackers.

The trick petals are at the bottom of the flower, forming a kind of envelope. Squeezed tightly inside are the stamens and pistil, ready to spring out. All that's needed is the weight of a heavy bumblebee. Then, poof! The petal envelope explodes open.

Ten stamens and the pistil explode around the bee. The pistil picks up pollen the bee is carrying from another broom flower, and the stamens dust the bee with a new load. Sometimes the flower parts pin the bee to the flower. But the bee doesn't seem to mind. It flies quickly from flower to flower, collecting lots of pollen for itself while at the same time pollinating the flowers. Poof, poof, poof!

This Scotch broom flower is waiting to explode. The weight of a landing bumblebee will pop open the flower's petal envelope.

Dodging the Blow

Alfalfa flowers also explode. Some honeybees have learned to steal the nectar without exploding the flower and getting bopped.

The Grass Pink's Fake Beard

It takes a lot of energy for plants to make pollen. Several flowers save energy by giving their pollinators make-believe pollen. One such faker is the grass pink (*Calopogon tuberosus*), a beautiful wild orchid of the eastern United States and Canada. Its upper petal appears to brim with golden pollen. But the "pollen" is really a golden beard of hairs.

As soon as a bee lands on the beard, the petal flops forward, tipping the bee on its back. As the bee slides down the center of the flower, packets of real pollen stick to its back. When the bee visits another grass pink, it is again fooled by the golden beard. This time,

Hungry bees visit the golden beards of grass pinks looking for pollen. But the bees find only hairs and a trick petal that sends them on a wild pollinating ride.

48

as the bee falls and slides, it first deposits its pollen load on the stigma and then picks up some more.

The Christmas Tree of Australia

Plants can't survive without water. For plants living where it is dry, getting enough water can be a major problem. But not for the Christmas tree (*Nuytsia floribunda*), one of the tallest plants of western Australia. Christmas arrives in western Australia during the hottest part of their summer. Most plants have died

The Christmas tree of Australia blooms in glorious color at the hottest time of the year by stealing water from plants all around it.

or wilted under the hot sun. But that's when the Christmas tree blooms, all aglow with brilliant orange-gold flowers.

What is its secret? It steals water from all of the other plants around it. Its roots develop **suckers** that plug into the roots of other plants. Before the other plants can use the water they have collected, the Christmas tree diverts it into its own roots. Any victim will do. The Christmas tree's roots make thousands of connections—to trees, grasses, and even roses and carrots.

The Squirting Cucumber

The squirting cucumber (*Ecballium elaterium*) is a common weed of the Mediterranean area. Its small green fruits hang, like hairy eggs, at the top of long stalks. As the fruits ripen, they get ready to blast off. The fruits fill with slimy juice. Inside each fruit, the pressure builds and builds.

When jarred by an animal, the fruit explodes from its stalk. It shoots through the air like a rocket. A trail of slime and seeds squirt out behind. The fruit and seeds travel up to 20 feet (6 m) away from the parent plant. In their new places, the seeds stand an excellent chance of growing into new plants.

The Monkey's Dinner Bell Tree

Bang! Bang! Sounds like gunshots echo through forests in South America. Monkeys come running. They know the bangs mean the big seedpods of the monkey's dinner bell tree (*Hura crepitans*) are exploding. The seeds are a favorite food of the monkeys.

The seedpods, shaped like little pumpkins, explode when they become ripe and dry. They are about the size of oranges, and actually split into sections like oranges when they explode. Each section holds a flat, round seed. The pieces of shell and seeds fly 40 feet (12 m) or more through the air. The monkeys

The fruits of squirting cucumber plants rocket through the air, squirting behind them a trail of slime and seeds.

51

don't eat all of the seeds, and some of them grow into new trees in their new homes.

Tricks and clever devices help plants survive from day to day and into the future. As the world continues to change, plants will slowly develop new and perhaps even trickier ways. The plants with the most successful strategies will be the ones most likely to survive.

These large seedpods of the monkey's dinner bell tree will explode with a bang, sending their orange sections through the forest.

Glossary

adapt—to change over time to better survive in a particular environment

anther—the top part of a flower's stamen, which produces pollen

aphid—a tiny insect that sucks juice from plants

cross-pollination—the transfer of pollen from the male part of one plant to the female part of another plant of the same species

dung—animal droppings

fertilize—to unite a male cell and a female cell to form a new individual

filament—the thin stalk that supports a flower's anther

insecticide—a chemical that harms or kills insects

leaflet—a small section of a leaf that is divided into many parts

mammal—a warm-blooded animal with a backbone; the females have milk glands to feed their young.

mimic—to imitate

nectar—a sweet liquid made by flowers that attracts pollinators

nectar or **honey guide**—a marking on flowers that guides insects to nectar and pollen and helps ensure pollination

ovule—the female cell of plants, which develops into a seed after fertilization

pistil—the female reproductive part of a flower, where the seeds are developed

pollen—the male cells of flowers and other seed plants, and a food that attracts pollinators

pollination—the transfer of pollen from a male part of a plant to a female part

pollinator—an animal that assists in pollination

seed—a fertilized ovule from which a new plant grows

species—a group of closely related animals or plants that can breed with one another

spore—a tiny specialized structure that can give rise to a new plant

stamen—the male reproductive part of a flower, where the pollen is produced

stigma—the part of a flower that receives the pollen grains

style—the stalk that supports a flower's stigma

sucker—a short shoot or branch that develops from plant roots

ultraviolet—a color not visible to humans, but visible to many insects

To Find
Out More

Books

Johnson, Sylvia A. *Roses Red, Violets Blue: Why Flowers Have Colors*. Minneapolis, MN: Lerner Publication Co., 1991.

Kalman, Bobbie. *What Is a Plant?* New York: Crabtree Publishing Co., 2000.

Kneidel, Sally. *Skunk Cabbage, Sundew Plants and Strangler Figs: The Strangest Plants on Earth*. New York: John Wiley & Sons, 2001.

Lasky, Kathryn. *The Most Beautiful Roof in the World: Exploring the Rainforest Canopy*. San Diego: Gulliver Green/Harcourt Brace & Company, 1997.

Taylor, Barbara. *Incredible Plants.* New York: DK Publishing, Inc., 1997.

Tesar, Jenny. *Green Plants.* Woodbridge, Conn.: Blackbirch Press, 1993.

Video

The Private Life of Plants. British Broadcasting Corporation, Inc., 1995.

Organizations and Online Sites

B-EYE: See the World through the Eyes of a Honey Bee
http://cvs.anu.edu.au/andy/beye/beyehome.html
This site lets you see how a bee views the world.

Encyclopedia Smithsonian: Botany
http://www.si.edu/resource/faq/nmnh/botany.htm
This site takes you to orchid and other botany exhibits at the Smithsonian Institution.

Hart Prairie Preserve Station W: Pollination Ecology
http://www.infomagic.net/~tnc/tours/stationw.htm
This site shows the scarlet gilia in Arizona that changes color when its pollinator changes.

International Aroid Society, Inc.
http://www.aroid.org
This site has many pictures and details about arums, stinky and otherwise.

Plants National Database
http://plants.usda.gov
This United States government site lets you search for photos and descriptions of thousands of plants.

Royal Botanic Gardens, Kew: Education
http://www.rbgkew.org.uk/education/index.html
This site from London has information sheets on rattans and other interesting plants.

Wayne's Word: A Newsletter of Natural History Trivia
http://waynesword.palomar.edu/wayne.htm
This site by Professor Wayne P. Armstrong is a treasure trove of plant science, photos, and lore.

Zoologisk Museum Pollination Exhibition:
"The Birds and the Bees"
http://www.toyen.uio.no/zoomus/engelsk/bird_and_bees/introduction.html
At this site, you can see wonderful close-up photos of insects pollinating flowers.

A Note on Sources

The first thing I did to research this book was watch *The Private Life of Plants*, a fascinating six-part video series by well-known nature writer and filmmaker David Attenborough. In these videos, he took me around the world to see hundreds of amazing plants up close and to share their stories firsthand. The companion book to the series was also a wonderful resource.

I also scoured the library for books and journal articles by scientists on plant life, mimicry, unusual plants, and relationships between plants and animals. Some of the most helpful books were *The Natural History of Pollination* by Michael Proctor, Peter Yeo, and Andrew Lack; *The Sex Life of Flowers* by Bastiaan Meeuse and Sean Morris; and *The Lives of Plants* by Doris M. Stone.

Early in the book's planning, a small item on the Internet about a flower changing color aroused my curiosity. Much additional research on the Internet and in scientific journals led to the chapter on the amazing ways flowers change color. Most useful was the research of Professor Martha R. Weiss.

Of special help on the Internet was a section dedicated to stinking flowers on the *Wayne's Word* web site of Professor Wayne P. Armstrong.

Index

Numbers in *italics* indicate illustrations.

About the Author

Plants have been an important part of Janet Halfmann's life as long as she can remember. As a child, she squatted in the farm field with her dad to check how much the corn or wheat had grown. Now she spends much of her spare time in her garden or watching the goings-on in the huge old tree outside her home-office window.

This is Halfmann's second book on plants. She is also the author of *Peanuts*, part of a Let's Investigate Agriculture series. Halfmann has written several children's nonfiction books, many of them on nature. They include a six-book series on wildlife habitats and several books on the lives and behavior of insects and spiders.

Before Halfmann became a full-time freelance children's writer, she was a newspaper reporter, children's magazine editor, and children's activity book writer and editor. Halfmann shares her never-ending fascination at the wonders of nature with her husband, four children, and grandchild.